U0108628

方智敏 編著

▲ 新手入廚系列

焗爐美食

前言

香港寸金尺土，地方有限，未必能放置大型焗爐（烤箱），令到一群熱愛烘焙的粉絲只能望門興嘆，不能盡情釋放烹飪的熱情，大煮特煮。幸運地，隨着一股製造甜品熱潮的強勢下，許多生產小型家庭電器的廠商洞悉商機，製造出許多輕巧時尚的小焗爐，通稱「小焗爐」，讓我們這群烘焙愛好者繼續開心有「焗」。

市面上的小焗爐，有些帶有微波功能。不過我卻愛用傳統功能，特別是有上、下發熱線或石英管作導熱媒體的。最好選用可以隨意選擇能關掉上火、下火、全開或全熄的焗爐，因為能靈活運用和處理食品的效果。

現今講求環保和飲食健康，利用焗爐作烹調工具，清潔、衛生又方便。它擁有翻熱、烹調，甚至附有蒸氣的功能，能配合少油、少炸等健康烹調的要求，用途多而廣泛，適應現代人的需要。

由於家居地方有限，都市生活繁忙，所以書中食譜以簡單、容易處理和隨時能購買的材料為依歸，有些甚至預先處理便可以輕鬆烹煮，隨時可以享受美食。

目錄

選購心水焗爐

Choosing Oven

焗爐尺碼有大有小，有座地式（全座爐身，以煤氣或電作能源）、座枱式（爐身有如紙箱般大小，安放在枱上）、鑲嵌式（爐身比座枱式大，可安在桌面上或牆壁）和座架式（典型小烤箱，尺碼小，只可作有限度烤焗和翻熱用途）。

不同大小的焗爐

1. 座地式和鑲嵌式焗爐屬高價爐，有火力和時間調控，兼有一按式功能圖示（一般會預設某些食物的烹調時間，如曲奇、烤雞或燒豬排等），有些會有上火和下火，有些則只有底火，焗爐空間大，可烤焗大量製品或大尺碼食物。

2. 座枱式焗爐適合一般家庭應用，尺碼適中，可烘烤一般大小的食物，主要利用發熱線或石英管發熱烤焗食物，它亦分有火力指引或圖像示意烹調，並設有上火、下火、全火或全熄火的功能，有些高檔爐還附有風扇把熱力吹遍全爐，讓製品效果更理想。

3. 座架式小焗爐，爐的容量很小，沒有清晰火力劃分，只有統一火力，只適合翻熱和烤製輕巧美食，不能長時間烘烤，否則會容易造成爐溫過熱的弊病。

焗爐應用小貼士

1. 使用焗爐前，先取出爐中所有雜物。

2. 烹調前，預熱焗爐，讓溫度平均，才方可把食物入爐。

3. 烹調時，應留意食物的情況和變化，按需要來遮蔽食物，以免燒焦。

4. 如果烹調雞、豬排、牛排或羊排等有油脂飛濺出來的東西，可在爐內墊上錫紙，方便烹調後清洗處理。

5. 剛烹調完食物的焗爐，可借助餘溫翻熱餅食。

6. 焗爐用後，先切斷電源，打開爐門，待涼後可用少許清潔劑清潔，並用清水抹乾淨。

香草沙律多士 / Salad Toast with Herbs

⚪⚪ 材料 | Ingredients

法包 1 條
蟹子適量（裝飾）

1pc French bread
crab roe as needed (for dressing)

⚪⚪ 香草牛油 | Herb butter

牛油 50 克
雜香草碎 20 克
蒜蓉 2 茶匙

50g butter
20g ground mixed herbs
2 tsps minced garlic

雜果沙律 | Mixed fruit salad

鮮雜果 100 克	100g mixed fruits
沙律醬 100 克	100g mayonnaise
煉奶 1 茶匙	1 tsp condensed milk

做法 | Method

1. 法包切片；香草牛油拌勻。
2. 雜果沙律材料拌勻，置冰箱中冷凍 1 小時，備用。
3. 把香草牛油塗在法包上。
4. 小焗爐預熱 3-5 分鐘，放入香草牛油法包焗 5-8 分鐘或至金黃。
5. 取出，上面放雜果沙律即可。

1. Slice French bread and mix herb butter well.
2. Stir the ingredients of mixed fruit salad well, put in the freezer to cool for 1 hour, set aside.
3. Spread herb butter onto French bread.
4. Preheat mini oven for 3-5 minutes, bake French bread with herb butter in oven for 5-8 minutes until golden yellow.
5. Take out and top with mixed fruit salad, serve.

> 入廚貼士 | Cooking Tips
> - 香草牛油法包必須趁熱享用，否則會變得很硬，不好吃。
> - Serve French bread with herb butter hot, otherwise the bread will become stiff when cool and not delicious any more.

蜜糖葡萄麵包飛碟

Honey Raisins Toastie

1~2 人
Serves 1~2

10~15 分鐘
10~15 minutes

⬭ 材料 | Ingredients

三文治麵包 2-4 片
葡萄乾 50 克
蜜桃粒 50 克
牛油熔液 50 毫升
蜜糖 1-2 茶匙
清水 1/3 杯

2-4 slices sandwich bread
50g raisins
50g peach dice
50g melted butter
1-2 tsps honey
1/3 cup water

⬭ 做法 | Method

1. 三文治麵包切去邊，備用。
2. 燒熱鍋，葡萄乾與清水同置煲中煮至汁液收乾，加入蜜桃粒和蜜糖拌勻。
3. 在模具上塗點牛油熔液，放上三文治麵包，再放入蜜糖葡萄和蜜桃粒，蓋上另一片麵包，壓實。
4. 小焗爐預熱 3-5 分鐘，放進麵包烤 5-8 分鐘，即成。

1. Cut the edges of sandwich bread, set aside.
2. Put raisins and water in a pot and cook until sauce thickens, add peach dice and honey, stir well.
3. Spread some melted butter onto the mould, put in the sandwich bread. Add honey raisins and peach dice. Cover with another piece of bread, press hard.
4. Preheat mini oven for 3-5 minutes and toast bread for 5-8 minutes, serve.

焦糖燒多士

Roasted Toast with Caramel

材料 | Ingredients

三文治麵包 2-4 片
雞蛋 2 隻
牛油 50 克

2-4 slices sandwich bread
2 eggs
50g butter

伴食 | Condiments

金絲糖膠 50 克
葡萄乾 20 克

50g syrup
20g raisins

做法 | Method

1. 三文治去邊，切成 2.5 厘米 × 2.5 厘米，備用。
2. 雞蛋在碗中打散，放入麵包小塊。
3. 小焗爐預熱 3-5 分鐘，牛油塗於焗盤上，再放上雞蛋麵包焗 3-5 分鐘，翻轉，再焗 3-5 分鐘或至金黃。
4. 伴以葡萄乾和金絲糖膠。

1. Cut edges of sandwiches, cut into 2.5cm × 2.5cm squares, set aside.
2. Whisk eggs in a bowl, add bread pieces, set aside.
3. Preheat mini oven for 3-5 minutes, spread butter onto oven tray. Put bread pieces onto oven tray and bake for 3-5 minutes. Turn over and bake for another 3-5 minutes or until golden yellow.
4. Serve with raisins and syrup.

入廚貼士 | Cooking Tips
- 可任用何乾果代替葡萄乾。
- The raisins can be replaced by any kind of dried fruits.

鳳梨酥

Pineapple Pastry

2~4 人
Serves 2~4

15~20 分鐘
15~20 minutes

材料 | Ingredients

鳳梨餡 540 克	540g pineapple fillings
低筋麵粉 220 克	220g cake flour
酥油 150 克	150g shortening
糖 50 克	50g sugar
奶粉 30 克	30g milk powder
芝士粉 15 克	15g cheese powder
雞蛋 1 隻	1 egg

做法 | Method

1. 酥油和糖放碗中攪拌至淡黃色，雞蛋分次打入拌勻。

2. 所有粉料一同過篩，倒入酥油混合物內拌勻，搓揉成粉糰。

3. 粉糰分成 18 份（每份重約 30 克），包上餡料，壓在鳳梨酥模，壓平。

4. 小焗爐預熱，將粉糰放入焗盤上，用 180℃焗 10 分鐘或至金黃，翻轉再焗 8 分鐘，即成。

1. Mix shortening and sugar in a bowl and stir until pale yellow in color, add egg in several times, stir well.

2. Sieve different kinds of powder together. Mix into shortening mixture and stir well, knead into a dough.

3. Divide dough into 18 portions, each weighs about 30g. Wrap in fillings and put into pineapple mould, press flat.

4. Preheat mini oven, put the molded dough onto oven tray and bake at 180℃ for 10 minutes or till golden yellow. Turn over and bake for 8 minutes, serve.

入廚貼士 | Cooking Tips
- 市面上有不同水果餡現貨可選擇。
- There are many kinds of fruit fillings available in the market, you can choose according to your preference.

4~6 人
Serves 4~6

15~20 分鐘
15~20 minutes

菠蘿肉腸披薩

Pineapple and Sausage Pizza

⊂⊃ 麵糰 | Dough

高筋麵粉 100 克
牛油 4 克
砂糖 3 克
酵母（依士）2 克
鹽 1 克
水 64 毫升

100g high gluten flour
4g butter
3g sugar
2g yeast
1g salt
64ml water

餡料 | Fillings

蒙沙拉芝士 100 克	100g mozzarella cheese
沙樂美腸 30 克	30g salami
菠蘿 2-3 片（切粒）	2-3 slices pineapple (dice)
茄汁 3 湯匙	3 tbsps ketchup

做法 | Method

1. 先將麵糰乾材料混合，再拌入水搓揉成糰，再加入牛油搓至光滑，封上保鮮紙，置旁待發酵 40 分鐘。
2. 取麵糰碾成圓麵皮，放在批碟上，刺孔，再作第二次發酵 15 分鐘。
3. 焗爐預熱至 3-5 分鐘，放入薄餅焗 5-8 分鐘，取出。
4. 按序放上茄汁、沙樂美腸、菠蘿和芝士，回爐焗 10-12 分鐘或至金黃便可。

1. Mix dough ingredients in a bowl, add water and knead into a dough. Add butter and knead till smooth, wrap with cling wrap. Set aside and let ferment for 40 minutes.
2. Roll the dough into circular shape, put onto a pie plate, poke holes. Let ferment again for 15 minutes.
3. Preheat mini oven for 3-5 minutes, bake pizza for 5-8 minutes, take out.
4. Top with ketchup, salami, pineapple and cheese in sequence. Bake for 10 -12 minutes again or until golden yellow, serve.

入廚貼士 | Cooking Tips

- 麵糰分次發酵，可讓披薩更鬆軟，效果更理想。
- Ferment the dough twice will make the pizza more tender and obtain better effect.

2~4 人
Serves 2~4

25~30 分鐘
25~30 minutes

芝士西蘭花焗薯

Baked Potatoes with Cheddar Cheese and Broccoli

材料 | Ingredients

美國焗薯（中）2 個
西蘭花 50 克
濃味車打芝士 4 片
即食煙肉碎 1-2 湯匙

2 American potatoes (medium)
50g broccoli
4 slices cheddar cheese
1-2 tbsps instant chopped bacon

⊂⊃⊃ 調味料 | Seasonings

上湯 1 杯　　　　　　　1 cup stock
鹽 1/4 茶匙　　　　　　1/4 tsp salt
胡椒粉適量　　　　　　Pinch of pepper

⊂⊃⊃ 做法 | Method

1. 焗薯洗淨；燒滾一鍋水，放入焗薯煮 15 分鐘至八成熟，剖開備用。

2. 西蘭花洗淨，切成小朵，用上湯焯煮 3-4 分鐘，過冷，備用。

3. 小焗爐預熱 2-3 分鐘，把焗薯放在焗盤上，按序放上西蘭花、芝士和煙肉碎焗 8-10 分鐘即成。

1. Rinse potatoes thoroughly. Boil a pot of water, add potatoes and boil for 15 minutes until 80% done, cut open and set aside.

2. Rinse brocooli and cut into pieces. Scald broccoli in the stock for 3-4 minutes, rinse with cold water, set aside.

3. Preheat mini oven for 2-3 minutes, put potatoes onto oven tray. Add broccoli, cheese and chopped bacon in sequence, bake for 8-10 minutes, serve.

入廚貼士 | Cooking Tips

- 想焗薯的味道集中，可以把原隻焗薯用叉子刺孔，裏上錫紙後放入焗爐焗熟，時間會長一點，不過味道會更濃郁。

- To make the potato tastier, you may prick the whole potato with a fork, wrap it with aluminum foil and bake in oven until done. This may take longer time but the taste will be richer.

香焗薯皮

Baked Potato Skins

材料 | Ingredients

新薯 3-4 個
煙肉片 2 片
蒙沙拉芝士（披薩芝士）100 克
洋蔥 50 克
蔥 1 條

3-4pcs new potatoes
2 slices bacon slices
100g Mozzarella cheese
(pizza cheese)
50g onion
1 stalk spring onion

2~4 人
Serves 2~4

25~30 分鐘
25~30 minutes

⦿ 調味料 | Seasonings

蒜鹽 1/2 茶匙
Garlic salt 1/2 tsp

⦿ 做法 | Method

1. 煮滾一鍋水，加入新薯煮 8 分鐘或放入小焗爐焗 10 分鐘，至約八成熟，取出對半剖開，小心挖去薯肉成小碗形。

2. 煙肉切碎，放入小焗爐焗至香脆。蔥切粒，洋蔥切碎。

3. 小焗爐預熱 2-3 分鐘，在薯皮分別放入蒜鹽、煙肉碎、蔥粒和洋蔥碎，再撒上芝士碎焗 5 分鐘，至芝士融化而呈金黃色便可。

1. Boil a pot of water, cook new potatoes for 8 minutes or bake in oven for 10 minutes until 80% done. Take out and cut open, remove flesh carefully to make a bowl-like shape.

2. Mince bacon, bake in oven until crispy. Dice spring onion and mince onion.

3. Preheat oven for 2-3 minutes, add garlic salt, chopped bacon, spring onion dice and chopped onion onto potato skin, sprinkle with grated cheese. Bake in oven for 5 minutes until cheese melts and in golden color, serve.

入廚貼士 | Cooking Tips

- 市面也有現成薯皮售賣，不過自己親手弄的薯皮味道會甘香點。
- Ready-to-serve potato skins are available in market but DIY potato peels are more tasty.

蒜香雜菜

Roasted Vegetables with Garlic

材料 | Ingredients

三色椒 100 克	100g green, yellow and red pepper
任何時菜 50 克	50g seasonal vegetables
鮮冬菇 3-4 朵	3-4pcs fresh black mushrooms
茄子 1 個	1 eggplant
橄欖油 / 牛油熔液 1 湯匙	1 tbsp olive oil / melted butter
蒜蓉 1 茶匙	1 tsp minced garlic

調味料 | Seasonings

燒汁（任何味道）1 茶匙	1 tsp Teriyaki sauce (any flavor)
蒜鹽 1 茶匙	1 tsp garlic salt

做法 | Method

1. 把雜菜洗淨，切件。
2. 將雜菜件放在錫紙上，再放上蒜蓉和橄欖油，封好。
3. 焗爐預熱 2-3 分鐘，放入雜菜焗 8-10 分鐘，取出加入調味便可。

1. Rinse vegetables thoroughly, cut into pieces.
2. Put vegetable pieces onto a piece of aluminum foil, add minced garlic and olive oil, seal.
3. Preheat mini oven for 2-3 minutes, bake vegetables into oven for 8-10 minutes. Take out and add seasonings, serve.

入廚貼士 | Cooking Tips

- 三色椒可以直接放入焗爐燒至軟腍或外皮焦腍，脫去外皮，拌以意大利黑醋、胡椒粒和橄欖油享用。
- You may roast green, yellow and red peppers in oven till soft or burnt on the surface. Remove the peel and serve with balsamic vinegar, peppercorn and olive oil.

肉碎焗釀大蘑菇

Baked Portobello Mushrooms
Stuffed with Minced Pork

2~4 人
Serves 2~4

18~20 分鐘
18~20 minutes

材料 | Ingredients

澳洲大啡蘑菇 2 朵	2 Portobello mushrooms
免治豬肉 100 克	100g minced pork
紅椒粉適量（裝飾）	Some red pepper powder (for dressing)
番茜碎適量（裝飾）	Some chopped parsley (for dressing)

醃料 | Marinade

雞蛋 1 隻	1 egg
糖 1 茶匙	1 tsp sugar
醬油 1 茶匙	1 tsp soy sauce
生粉 1 茶匙	1 tsp cornstarch
油 1 茶匙	1 tsp oil
蒜蓉 1 茶匙	1 tsp minced garlic
鹽 1/2 茶匙	1/2 tsp salt

做法 | Method

1. 免治豬肉與醃料拌勻，置冰箱中冷凍 30 分鐘。
2. 大蘑菇洗擦乾淨，釀入免治豬肉，撒點紅椒粉。
3. 焗爐預熱 2-3 分鐘，大蘑菇放入焗爐焗 15-18 分鐘或至熟。

1. Mix minced pork with marinade well, put into freezer for 30 minutes.
2. Rinse and wipe portobello mushrooms thoroughly, stuff with minced pork. Sprinkle with red pepper powder.
3. Preheat mini oven for 2-3 minutes, bake portobello mushrooms for 15-18 minutes or until done.

入廚貼士 | Cooking Tips

- 大蘑菇使用前才用清水略沖，不宜浸濕，否則蘑菇會變壞。
- Rinse mushrooms only before use, soaking is not recommended or else the mushrooms will rot.

2~3 人
Serves 2~3

8~10 分鐘
8~10 minutes

Roasted Prawns
with Salt and Chili

海鹽燒辣味蝦

◯◯◯ 材料 | Ingredients

大花蝦 4-6 隻	4-6 king prawns

◯◯◯ 醃料 | Marinade

糖 2 茶匙	2 tsps sugar
鮮露 1 茶匙	1 tsp seasoning sauce
酒 1 茶匙	1 tsp wine
海鹽 1/2 茶匙	1/2 tsp sea salt
味椒鹽 1/4 茶匙	1/4 tsp salt pepper seasoning

◯◯◯ 伴食 | Condiments

檸檬 3-4 角	3-4 lemon wedges

◯◯◯ 做法 | Method

1. 花蝦洗淨，挑去蝦腸，抹乾備用。
2. 醃料拌勻，放入蝦肉中醃 5 分鐘
3. 焗爐預熱 2-3 分鐘，放入花蝦焗 8-10 分鐘或至熟，取出與檸角同食。

1. Rinse king prawn, devein, wipe dry, set aside.
2. Stir marinade well and mix with prawns, leave for 5 minutes.
3. Preheat oven for 2-3 minutes, bake prawns for 8-10 minutes or until done. Take out and serve with lemon wedges.

入廚貼士 | Cooking Tips

可以在蝦背剖開一點，使蝦肉更加入味和比較快熟。
Cut a shallow slit at the back of prawns, make them get done quicker and flavor being absorbed easier.

芝士煙肉燒蠔

Roasted Oysters with Cheese and Bacon

材料 | Ingredients

連殼生蠔 4 隻
煙腩肉 3-4 片
瑞士芝士碎 50 克
洋蔥碎 30 克

4 pcs oysters in shells
3-4 slices smoked bacons
50g Emmenthalar cheese
30g chopped onion

2~4 人
Serves 2~4

25 分鐘
25 minutes

調味料 | Seasonings

海鹽 1/2 茶匙
黑胡椒碎 1 茶匙
Tabasco 辣椒汁 1 茶匙

1/2 tsp sea salt
1 tsp ground black pepper
1 tsp tabasco

做法 | Method

1. 生蠔洗擦乾淨，備用。
2. 煙腩肉切碎，預熱焗爐，煙腩肉放焗爐中焗至香脆，需時 10 分鐘，取出備用。
3. 煙腩肉的油加入洋蔥碎中，放焗爐焗 5 分鐘。
4. 生蠔分別放上調味料、煙肉碎和洋蔥碎，再撒上芝士。
5. 焗爐預熱 2-3 分鐘，放入生蠔焗 8-10 分鐘或至熟，即可。

1. Rinse and wipe oysters thoroughly, set aside.
2. Mince smoked bacon and bake in oven about 10 minutes or until crispy, take out and set aside.
3. Add oil from smoked bacon to chopped onion and bake in oven for 5 minutes.
4. Add seasonings, chopped bacon and onion to oysters, sprinkle cheese over.
5. Preheat oven for 2-3 minutes and bake oysters in oven for 8-10 minutes or until done, serve.

入廚貼士 | Cooking Tips

- 煙肉表面必須焗至乾身和金黃色，效果才理想。
- For ideal effect, the surface of bacon should be baked until dry and golden in color.

2~4 人
Serves 2~4

10~12 分鐘
10~12 minutes

鹽燒翡翠螺

Salt-roasted Jade Spiral Shells

主菜 ● 海鮮
Main Dish
Seafood

材料 | Ingredients

翡翠螺 4-8 隻	4-8pcs jade spiral shells
粗鹽 1 包	1 pack coarse salt
清酒 1-2 湯匙	1-2 tbsps sake

調味料 | Seasonings

味椒鹽適量　　　　　　Some salt pepper seasoning

做法 | Method

1. 翡翠螺擦洗乾淨，備用。
2. 焗爐預熱 2-3 分鐘，用錫紙墊在盤上，放上粗鹽，焗熱。
3. 取出，翡翠螺插在粗鹽中，回爐焗 5-8 分鐘，再取出，淋上清酒，回爐再焗 2-3 分鐘。
4. 享用時可灑點味椒鹽調味。

1. Wipe and rinse jade spiral shells thoroughly, set aside.
2. Preheat oven for 2-3 minutes, lay a sheet of aluminum foil on the tray, cover with coarse salt, bake until hot.
3. Take out, plunge jade spiral shells into coarse salt, bake in oven for 5-8 minutes. Take out again and sprinkle with sake, bake in oven again for 2-3 minutes.
4. Sprinkle with salt pepper seasonings when serve.

入廚貼士 | Cooking Tips

烤螺必須完全熟透，否則容易引起腸胃不適。
Bake spiral shells till well done, otherwise stomach upset may occur.

墨魚荔蓉盒

Cuttlefish and Taro Paste Box

芋頭 200 克
墨魚膠 100 克
固體菜油 / 豬油 50 克
牛油熔液 50 克（掃面）
澄麵 1-2 湯匙

200g taro
100g cuttlefish paste
50g margarine /
solid lard
50g melt butter
(for surface brushing)
1-2 tbsps wheat starch

2~4 人
Serves 2~4

25~30 分鐘
25~30 minutes

墨魚調味料 | Seasonings for cuttlefish paste

蛋白 1 隻	1 egg white
生粉 1 茶匙	1 tsp cornstarch
鹽 1/2 茶匙	1/2 tsp salt
糖 1/2 茶匙	1/2 tsp sugar
胡椒粉適量	Pinch of pepper
麻油適量	Some sesame oil

芋頭調味料 | Seasonings for taro

鹽 1/2 茶匙	1/2 tsp salt
五香粉 1/2 茶匙	1/2 tsp five-spice powder

做法 | Method

1. 墨魚膠與調味料放碗中拌勻，放入冰箱中冷凍 30 分鐘，備用。
2. 芋頭去皮，以大火蒸熟約 10-15 分鐘，取出，趁熱拌入調味料和固體菜油（乳瑪琳）或豬油搓爛成芋蓉。
3. 墨魚膠分成 10 份；芋蓉也分成 10 等份，每份芋蓉包上 1 份墨魚膠，做成盒形。
4. 焗爐預熱 2-3 分鐘，每個芋蓉盒塗上牛油熔液，放入焗爐焗 10-15 分鐘至熟透，即成。

1. Mix cuttlefish paste with seasonings well, put into freezer for 30 minutes, set aside.
2. Peel taro, steam over high heat for 10-15 minutes or till done. Take out taro, mix with seasonings and vegetable oil when it is hot, mash into taro paste.
3. Divide cuttlefish paste and taro paste into 10 portions respectively, wrap each portion of cuttlefish paste with taro paste and make into box-shape.
4. Preheat oven for 2-3 minutes, brush each taro box with melted butter and bake for 10-15 minutes until done, serve.

入廚貼士 | Cooking Tips

- 可隨意改變自己喜愛的餡料。
- You may change the fillings as you like.

2~4 人
Serves 2~4

10~15 分鐘
10~15 minutes

雜燴脆籃

Seafood Mixtures in Crispy Basket

〇〇〇 材料 | Ingredients

Fillo 千層酥皮 5 張	5 sheets Fillo pastry
蝦仁粒 50 克	50g shrimp dice
豬瘦肉粒 50 克	50g lean pork dice
魷魚粒 50 克	50g squid dice
西芹粒 20 克	20g celery dice
甘筍片 20 克	20g carrot slices
牛油熔液 2-3 湯匙	2-3 tbsps melted butter
薑蓉 1 茶匙	1 tsps minced ginger
蒜蓉 1 茶匙	1 tsps minced garlic

醃料 | Marinade

糖 1 茶匙	1 tsp sugar
生粉 1 茶匙	1 tsp cornstarch
酒 1 茶匙	1 tsp wine
油 1 茶匙	1 tsp oil
鹽 1/2 茶匙	1/2 tsp salt
胡椒粉適量	Pinch of pepper
麻油適量	Some sesame oil

芡汁 | Sauce

上湯 3 湯匙	3 tbsps stock
生粉 1 茶匙	1 tsp cornstarch

做法 | Method

1. 蝦仁粒和魷魚粒用 1/2 份醃料拌勻；其餘醃料與瘦肉粒拌勻，分別汆水備用。
2. 熱鑊下油，爆香薑蓉和蒜蓉，放入豬瘦肉粒、蝦仁和魷魚粒炒透，加入雜菜粒和芡汁炒汁液至濃稠。
3. 焗爐預熱 2-3 分鐘，把每張 Fillo 千層酥皮掃上牛油，疊在一起，放在模具上，置焗爐焗至金黃香脆。
4. 將海鮮雜燴放在千層酥皮上即可。

1. Mix shrimp and squid dice with 1/2 of marinade, stir well; mix the remaining marinade with lean pork dice, stir well. Scald two mixtures separately, set aside.
2. Add oil to a heated wok, sauté minced ginger and garlic until fragrant, add lean pork dice, shrimp and squid dice, stir-fry until well done. Add vegetable dice and sauce and cook until sauce thickens.
3. Preheat oven for 2-3 minutes, brush each Fillo pastry with butter, stack together and put on the mould, bake in oven until golden and crispy.
4. Put seafood mixtures into puff pastry, serve.

日式燒汁烤鯖魚

Roasted Japanese Mackerels with Sauce

鯖魚 2-4 塊 （約 300 克）　2-4 Japanese mackerel pieces (about 300g)

北京大葱 50 克　50g Peking onion

油 1 湯匙　1 tbsp oil

⬤⬤ 醃料 | Marinade

檸檬汁 2 茶匙　2 tsps lemon juice

海鹽 1 茶匙　1 tsp sea salt

生粉 1 茶匙　1 tsp cornstarch

胡椒粉適量　Pinch of pepper

2~4 人
Serves 2~4

15~18 分鐘
15~18 minutes

⊚ 燒汁 | BBQ sauce

鰻魚燒汁 1 湯匙   1 tbsp eel sauce
味醂 2 茶匙   2 tsps Mirin (sweet sake)
保衛爾牛肉汁 1 茶匙   1 tsp Bovril beef sauce

⊚ 做法 | Method

1. 鯖魚洗淨，抹乾備用。
2. 醃料拌勻，塗抹在鯖魚上。
3. 北京大葱洗淨，切大塊。
4. 焗爐預熱 2-3 分鐘，把大葱放在焗盤上，再放上鯖魚塊，淋上油後焗 10-15 分鐘。
5. 取出鯖魚，把已調勻的燒汁塗抹在魚身，回爐焗至金黃，即成。

1. Rinse mackerels thoroughly, wipe dry, set aside.
2. Mix marinade well and brush onto mackerels.
3. Rinse Peking onion thoroughly and cut into big pieces.
4. Preheat oven for 2-3 minutes, put Peking onion onto oven tray and top with mackerels, sprinkle with oil and bake for 10-15 minutes.
5. Take out mackerels, brush well-mixed sauce onto the surface, bake in oven again until golden, serve.

入廚貼士 | Cooking Tips

- 把大葱墊底，可避免魚件黏底和燒焦，也可讓大葱在烘烤過程中釋出香味，被魚塊吸收，引發食慾。
- Putting Peking onion on the bottom can prevent mackerels being stuck onto the bottom and get burnt. Meanwhile, the fragrance released by Peking onion in the process of baking can be absorbed by mackerels and hence induce appetite.

2~4 人
Serves 2~4

20~25 分鐘
20~25 minutes

鮮茄燒魚

Roasted Grey Mullet with Tomato

 材料 | Ingredients

烏頭魚 1 條（約 450 克）
番茄 150 克
洋葱 100 克
橄欖油 1-2 湯匙

1 pc grey mullet
(about 450g)
150g tomatoes
100g onion
1-2 tbsps olive oil

◯◯◯ 醃料｜Marinade

百里香 3-4 條	3-4 stalks thyme
海鹽 1 茶匙	1 tsp sea salt
檸檬汁 1 茶匙	1 tsp lemon juice
糖 1 茶匙	1 tsp sugar
胡椒粉適量	Pinch of pepper

◯◯◯ 伴食｜Condiments

青檸 2-3 角	2-3 lime wedges
味椒鹽適量	Pinch of salt pepper
黑胡椒碎適量	Pinch of ground black pepper

◯◯◯ 做法｜Method

1. 番茄和洋蔥分別洗淨，切粒。
2. 烏頭魚劏洗乾淨，抹乾魚身。
3. 醃料調勻，塗抹在全魚，把百里香放在魚肚內。
4. 焗爐預熱 2-3 分鐘，在焗盤放上番茄粒和洋蔥粒，淋上橄欖油，再放上烏頭魚焗 15-20 分鐘至全熟。

1. Rinse tomatoes and onion and dice.
2. Rinse and gut grey mullet, wipe dry.
3. Mix marinade well, brush onto grey mullet, put thyme inside grey mullet.
4. Preheat oven for 2-3 minutes, put tomato and onion dice onto baking tray, sprinkle with olive oil, then top with grey mullet. Bake for 15-20 minutes until well done.

入廚貼士｜Cooking Tips

- 烏頭魚含油份比較重，適合烤焗。如果魚身太大，可切成數塊，方便烤焗，不過魚味會有所偏差。
- Grey mullet contains high content of oil and is suitable for baking. If the fish is too big, may cut into several pieces, however, the taste may vary a little bit.

醬汁燒九孔

Roasted Taiwan Abalones with Sauce

⊙⊙⊙ 材料 | Ingredients

台灣九孔鮑魚 6-8 隻
6-8 Taiwan abalones

⊙⊙ 芫荽水 | Coriander water

芫荽頭 2-3 個
清水 1/3 杯
2-3 pcs coriander roots
1/3 cup water

2~4 人
Serves 2~4

10~15 分鐘
10~15 minutes

燒汁 | BBQ Sauce

芫荽水 2 湯匙	2 tbsps coriander water
日式燒汁 1 湯匙	1 tbsp Japanese sauce
味醂 1 湯匙	1 tbsp mirin (sweet sake)
糖 1 茶匙	1 tsp sugar
雞粉 1/2 茶匙	1/2 tsp chicken powder
鮮露 1/2 茶匙	1/2 tsp seasoning sauce

做法 | Method

1. 芫荽水材料放煲中煮至濃縮為 2-3 湯匙份量，備用。
2. 把燒汁料放煲中煮滾，備用。
3. 九孔鮑魚洗擦乾淨，備用。
4. 焗爐預熱 2-3 分鐘，九孔鮑魚放焗爐焗 5-6 分鐘，取出掃上燒汁，回爐焗至鮑魚全熟。

1. Put ingredients of coriander water into a pot, cook until reduce to volume of 2-3 tablespoons, set aside.
2. Put sauce ingredients into the pot, bring to a boil and set aside.
3. Rinse and wipe abalones thoroughly, set aside.
4. Preheat oven for 2-3 minutes, put abalones into the oven and bake for 5-6 minutes. Take out and brush with the sauce, bake in oven again until well done.

入廚貼士 | Cooking Tips

- 鮑魚的裙邊容易貯藏細菌，必須用牙刷擦洗乾淨。
- Bacteria usually aggregate at the edges of abalones, so toothbrush must be used to brush and rinse thoroughly.

2-3 人
Serves 2-3

10-15 分鐘
10-15 minutes

香蒜牛油大花蝦

King Prawn with Garlic Butter

⊂⊃ 材料 | Ingredients

大花蝦 3-4 隻
3-4 king prawns

⊂⊃ 醃料 | Marinade

海鹽 1/2 茶匙
糖 1/2 茶匙
胡椒粉適量
1/2 tsp sea salt
1/2 tsp sugar
Pinch of pepper

⦿ 蒜蓉沙律醬 | Garlic dressing

牛油 20 克	20g butter
沙律醬 2 湯匙	2 tbsps mayonnaise
炸蒜蓉 1 湯匙	1 tbsp deep-fried minced garlic
蒜蓉 1 茶匙	1 tsp minced garlic
芥末醬 1 茶匙	1 tsp mustard

⦿ 做法 | Method

1. 大花蝦挑去腸，洗淨，背部剖開，加入醃料醃 5 分鐘。
2. 蒜蓉沙律醬拌勻，釀在大花蝦上。
3. 焗爐預熱 2-3 分鐘，放入大花蝦焗 10 分鐘或至熟，即成。

1. Devein king prawns and rinse thoroughly, slit on the back, add marinade and stir well, leave for 5 minutes.
2. Stir garlic dressing well, stuff onto king prawns.
3. Preheat oven for 2-3 minutes, bake king prawns for 10 minutes or until done, serve.

入廚貼士 | Cooking Tips

- 可先把大花蝦略煎後才釀入蒜蓉沙律醬，轉回焗爐烤焗，效果會更理想。
- To have a better effect, you may shallow-fry king prawns before stuffing with garlic dressing, and put back to the oven.

香辣醬油雞翼

Chili Chicken Wings

材料 | Ingredients

雞中翼 10-12 隻
洋葱絲 100 克

10-12 mid-joint chicken wings
100g onion shreds

醃料 | Marinade

糖 1 湯匙	1 tbsp sugar
橄欖油 1 湯匙	1 tbsp olive oil
海鹽 1 茶匙	1 tsp sea salt
鮮露 1 茶匙	1 tsp seasoning sauce
生粉 1 茶匙	1 tsp cornstarch
Tabasco 辣椒汁 1 茶匙	1 tsp tabasco
老抽 1/2 茶匙	1/2 tsp dark soy sauce
雜胡椒碎適量	Some mixed ground pepper

做法 | Method

1. 雞中翼解凍後，用少許檸檬汁拌勻，待 5 分鐘，沖洗乾淨。
2. 燒熱水一鍋，放入雞翼汆水，加入醃料拌勻，置雪櫃中冷凍 1-2 小時。
3. 焗爐預熱 2-3 分鐘，在焗盤上先放洋葱絲，再鋪上雞翼焗 10-12 分鐘。

1. Defrost chicken wings, mix well with some lemon juice, leave for 5 minutes, rinse thoroughly.
2. Boil a pot of water, scald chicken wings, add marinade and stir well, put in refrigerator for 1-2 hours.
3. Preheat oven for 2-3 minutes, put onion shreds onto baking tray, then top with chicken wings. Bake for 10-12 minutes, serve.

入廚貼士 | Cooking Tips

- 小焗爐的火力比較猛，烘烤時要多留意食物狀況，如果遇到表面或底部出現燒焦情況，可用錫紙蓋面或立即關掉上火或下火。
- Pay attention to food condition during baking as the heat of oven is rather strong. If the surface or bottom of the food is burnt, you may use aluminum foil to cover it or switch off the upper or lower heat immediately.

香檸燒雞鎚

Roasted Chicken Drumsticks with Lemon Slices

⊙⊙ 材料 | Ingredients

雞鎚 4-6 隻	4-6pcs chicken drumsticks
檸檬片 3-4 片	3-4 slices lemon slices
（裝飾或伴食）	(for decorationor serve together)

⊙⊙ 醃料 | Marinade

蛋白 1 隻	1 egg white
橄欖油 1 湯匙	1 tbsp olive oil
糖 2 茶匙	2 tsps sugar
鮮露 1 茶匙	1 tsp seasoning sauce
海鮮醬 1 茶匙	1 tsp seafood paste
生粉 1 茶匙	1 tsp cornstarch
黑胡椒碎 1 茶匙	1 tsp ground black pepper
酒 1 茶匙	1 tsp wine
鹽 1/2 茶匙	1/2 tsp salt
雞粉 1/2 茶匙	1/2 tsp chicken powder
蒜鹽 1/2 茶匙	1/2 tsp garlic salt

2~3 人
Serves 2~3

15~20 分鐘
15~20 minutes

⦿ 檸檬蜜糖水 | Lemon honey water

蜜糖 1 湯匙	1 tbsp honey
檸檬汁 1 茶匙	1 tsp lemon juice
熱水 1 茶匙	1 tsp hot water

⦿ 做法 | Method

1. 雞鎚解凍，用少許檸檬汁拌勻，待 5 分鐘，洗淨，抹乾水份。
2. 醃料拌勻，加入雞鎚內拌勻，置雪櫃中冷凍 1-2 小時。
3. 焗爐預熱 2-3 分鐘，放入雞鎚焗 13-15 分鐘。
4. 取出雞鎚，刺入竹籤確保沒有血水流出，拌勻並掃上檸檬蜜糖水，回爐焗 2-3 分鐘便可。

1. Defrost chicken thighs, mix well with some lemon juice, leave for 5 minutes, rinse and wipe dry.
2. Stir marinade well, add chicken thighs and mix well, put in refrigerator for 1-2 hours.
3. Preheat oven for 2-3 minutes, bake chicken thighs in oven for 13-15 minutes.
4. Take out chicken thighs, poke with a bamboo stick to ensure no bleeding. Mix lemon honey water well and brush onto chicken thighs, bake in oven again for 2-3 minutes, serve.

入廚貼士 | Cooking Tips

- 為縮減烹調時間，可在雞鎚厚肉部分用小刀�… 幾下，也比較容易入味。
- To shorten cooking time, you may slit the thick part of chicken thighs with a small knife. This will also shorten marinating time.

2~4 人
Serves 2~4

10~12 分鐘
10~12 minutes

香蒜燒和牛肉

Roasted Japanese Beef with Garlic

材料 | Ingredients

和牛肉 150 克	150g Japanese beef
蒜頭 2 粒	2 cloves garlic
橄欖油 1-2 湯匙	1-2 tbsps olive oil

醃料 | Marinade

橄欖油 1 湯匙	1 tbsp olive oil
鮮露 1 茶匙	1 tsp seasoning sauce
保衛爾牛肉汁 1 茶匙	1 tsp Bovril beef sauce
日式燒肉汁 1 茶匙	1 tsp Japanese sauce for BBQ meat
糖 1 茶匙	1 tsp sugar
生粉 1 茶匙	1 tsp cornstarch

做法 | Method

1. 蒜頭洗淨，切薄片，備用。

2. 醃料拌勻，塗抹在牛肉上，待 5 分鐘。

3. 焗爐預熱 2-3 分鐘，在焗盤上放上蒜片和牛肉，再淋上橄欖油，焗 5-6 分鐘，熄爐，待 1-2 分鐘取出即成。

1. Rinse garlic, cut into thin slices, set aside.

2. Stir marinade well, coat beef and leave for 5 minutes.

3. Preheat oven for 2-3 minutes, put garlic slices and beef onto a baking tray and sprinkle with olive oil. Bake for 5-6 minutes, switch off oven and leave for 1-2 minutes, serve.

入廚貼士 | Cooking Tips

- 挑選牛肉時，肉質的脂肪要多而均勻，烤出來的牛肉便會又嫩又滑。
- When choosing beef, pick the one with more fat and the fat being evenly distributed, the baked beef will be tender and smooth.

蜜汁燒香腸伴薯角
Honey Roasted Sausages with Potato Wedges

◯◯ 材料 | Ingredients

豬肉香腸 4-6 條
新薯 2-3 個
油 1-2 茶匙

4-6 pork sausages
2-3 new potatoes
1-2 tsps oil

◯◯ 調味料 | Seasonings

牛油 50 克
番荽碎 10 克
蒜鹽 1 茶匙

50g butter
10g chopped parsley
1 tsp garlic salt

2~4 人
Serves 2~4

25~30 分鐘
25~30 minutes

◯◯ **桂花蜜糖汁** | Sweet osmanthus honey sauce

蜜糖 / 麥芽糖 1 湯匙	1 tbsp honey / maltose
桂花糖 1 茶匙	1 tsp sweet osmanthus sugar
檸檬汁 1 茶匙	1 tsp lemon juice
熱水 1 茶匙	1 tsp hot water

◯◯ **做法** | Method

1. 新薯洗淨，切角，拌入調味料，放入已預熱的焗爐焗 8-10 分鐘或至熟。
2. 桂花蜜糖汁拌勻，備用。
3. 焗爐預熱 2-3 分鐘，把已塗油的肉腸放入焗爐烤 10-12 分鐘或至熟。
4. 取出豬肉腸，掃上桂花蜜糖汁，回爐焗 3-5 分鐘至甘香鬆脆。

1. Rinse potatoes and cut into wedges, mix with seasonings. Put into the preheated oven and bake for 8-10 minutes or until done.
2. Stir sweet osmanthus honey sauce well, set aside.
3. Preheat oven for 2-3 minutes, bake oil-coated sausages in oven for 10-12 minutes or until done.
4. Take out pork sausages and brush with sweet osmanthus honey sauce, bake in oven again for 3-5 minutes until crispy.

入廚貼士 | Cooking Tips
- 桂花蜜糖汁含糖份，容易令食物變得深色，所以必須小心掌管爐火，否則食物容易燒焦。
- Sweet osmanthus honey sauce contains sugar which will be easily get burnt, so be careful with the power of the oven and prevent the food from get burnt.

香燒羊扒

Roasted Lamb Chops with Herbs

◯◯◯ **材料** | Ingredients

羊扒 2-4 件	2-4 pcs lamb chops
橄欖油 1 湯匙	1 tbsp olive oil

◯◯◯ **調味料** | Seasonings

梳打餅碎 50 克	50g ground cracker
迷迭香 1 束	a bundle rosemary
千里香 1 束	a bundle thyme
番荽碎 1 湯匙	1 tbsp chopped parsley
牛油 1 湯匙	1 tbsp butter
蒜蓉 2 茶匙	2 tsps minced garlic

醃料｜Marinade

日式燒汁 1 湯匙	1 tbsp Japanese sauce
橄欖油 1 湯匙	1 tbsp olive oil
蒜鹽 1 茶匙	1 tsp garlic salt
糖 1 茶匙	1 tsp sugar
生粉 1 茶匙	1 tsp cornstarch

伴食｜Condiments

英式芥末	English style mustard

做法｜Method

1. 羊扒解凍，放入醃料醃約 5-10 分鐘。
2. 把調味料的香草洗淨，剁幼，加入其他材料拌勻，塗在羊扒上。
3. 燒熱鑊，把羊扒略煎，放在焗盤上。
4. 焗爐預熱 2-3 分鐘，放入羊扒焗 12-15 分鐘至全熟。

1. Defrost lamb chops, add marinade and leave for 5-10 minutes.
2. Rinse and shred herbs in seasonings, add other ingredients and stir well, brush onto lamb chops.
3. Heat a wok and slightly pan-fry lamb chops, put onto a baking tray.
4. Preheat oven for 2-3 minutes, bake lamb chops for 12-15 minutes till done.

入廚貼士｜Cooking Tips

- 羊扒先略煎，可把肉質鎖緊，保持肉汁於羊扒內。
- Slightly pan-fry lamb chops in advance can help to retain the meat juice.

芝麻醬汁燒牛扒

Roasted Steak with Sesame Sauce

2~4 人
Serves 2~4

15~18 分鐘
15~18 minutes

⦾ 材料｜Ingredients

西冷牛扒 2 件（約 300 克）	2 pcs sirloin steaks (about 300g)
芝麻 1-2 湯匙	1-2 tbsps sesame seeds

⦾ 醃料｜Marinade

日式燒汁 1-2 湯匙	1-2 tbsps Japanese sauce
味醂 2 茶匙	2 tsps mirin (sweet sake)
醬油 1 茶匙	1 tsp soy sauce
芝麻醬 1 茶匙	1 tsp sesame sauce
Tabasco 辣椒汁 1 茶匙	1 tsp tabasco

⦾ 做法｜Method

1. 芝麻浸水 30 分鐘，瀝乾。
2. 西冷牛扒用刀剁鬆身，放入醃料醃 15 分鐘。
3. 把牛扒黏上芝麻，燒熱油鑊，下牛扒略煎。
4. 焗爐預熱 2-3 分鐘，放入牛扒焗 8-12 分鐘至全熟。

1. Soak sesame seeds for 30 minutes, drain.
2. Tenderize sirloin steaks with a cleaver, add marinade and leave for 15 minutes.
3. Coat sirloin steaks with sesame. Heat a wok and pan-fry sirloin steaks slightly in a wok.
4. Preheat oven for 2-3 minutes, bake sirloin steaks for 8-12 minutes until done.

> 入廚貼士｜Cooking Tips
> - 芝麻浸水後才用，烤焗時不容易燒焦。
> - Soak sesame seeds before baking can avoid burning.

香草燒豬仔骨

Roasted Pork Ribs with Herbs

⌇⌇⌇ 材料 | Ingredients

豬仔骨 600 克
洋葱 100 克
西芹 50 克
甘筍 50 克

600g pork ribs
100g onion
50g celery
50g carrot

 醃料 | Marinade

黑胡椒碎 1 湯匙	1 tbsp ground black pepper
橄欖油 1 湯匙	1 tbsp olive oil
糖 2 茶匙	2 tsps sugar
茄汁 2 茶匙	2 tsps ketchup
蜜糖 2 茶匙	2 tsps honey
海鹽 1 茶匙	1 tsp sea salt
酒 1 茶匙	1 tsp wine
生粉 1 茶匙	1 tsp cornstarch
海鮮醬 1 茶匙	1 tsp seafood paste
Tabasco 辣椒汁 1 茶匙	1 tsp tabasco
雜香草 1 茶匙	1 tsp mixed herbs

做法 | Method

1. 洋葱洗淨，切圈；西芹和甘筍洗淨，切段。
2. 醃料拌勻，備用。
3. 豬仔骨解凍，在肉面剠紋，洗淨，抹乾，塗上醬料醃 30 分鐘。
4. 焗爐預熱 2-3 分鐘，按序放入洋葱、西芹條、甘筍條和豬仔骨焗 20-25 分鐘至全熟。

1. Rinse onion, cut into rings; rinse celery and carrot and cut into pieces.
2. Stir marinade well and set aside.
3. Defrost pork ribs and slit on surface, rinse and wipe dry. Spread marinade onto pork ribs and leave for 30 minutes.
4. Preheat oven for 2-3 minutes, put onion, celery sticks, carrot sticks and pork ribs in oven according in sequence, bake for 20-25 minutes to well done.

入廚貼士 | Cooking Tips

- 烤焗食物，特別是肉類，醃料的糖份會多一點，因為糖可以讓肉質軟滑和色澤光亮。
- The marinade for baking food will be rather sweet especially for meat because this will make the texture more tender and color brighter.

黑醋蜜糖燒豬腩

Roasted Pork Belly with Black Vinegar Honey Sauce

材料 | Ingredients

豬腩肉 300 克（煲湯後）
油 1-2 湯匙

300g pork belly
(cooked in soup)
1-2 tbsps oil

2~4 人
Serves 2~4

10~15 分鐘
10~15 minutes

⟨⟨⟨ 黑醋蜜糖汁 | Black vinegar honey sauce

黑醋 1 湯匙	1 tbsp black vinegar
蜜糖 1 湯匙	1 tbsp honey
鮮露 2 茶匙	2 tsps seasoning sauce
味醂 2 茶匙	2 tsps mirin (sweet sake)
糖 1 茶匙	1 tsp sugar
老抽 1 茶匙	1 tsp dark soy sauce

⟨⟨⟨ 做法 | Method

1. 把煲過湯的豬腩肉切成小件。
2. 黑醋蜜糖汁拌勻，置煲中煮滾，放入豬腩肉醃 20 分鐘。
3. 焗爐預熱 2-3 分鐘，把油淋在豬腩肉上，放入焗爐焗 10-15 分鐘，即成。

1. Cut boiled pork belly into small pieces.
2. Stir black vinegar honey sauce well, put in a pot and bring to a boil, add pork belly and leave for 20 minutes.
3. Preheat oven for 2-3 minutes, pour oil over pork belly and bake for 10-15 minutes, serve.

入廚貼士 | Cooking Tips

- 用煲湯豬肉的好處是不浪費，肉質也比較軟滑，只要加點味道便可以，省時省力。
- The advantage of using pork which has been used to boil soup is that it can reduce wastage and the texture of pork is tenderer. Just add some flavor will be fine enough. This is convenient and time-saving.

酥皮焗意粉

Baked Pasta with Pastry

材料 | Ingredients

酥皮 1/2 包	1/2 pack puff pastry
雞蛋 1 隻	1 egg

茄汁意大利粉 | Ketchup pasta

意大利粉 150 克（已煮熟）	150g pasta (cooked)
免治豬肉 50 克（汆水）	50g minced pork (scalded)
雜菜粒 2-3 湯匙（汆水）	2-3 tbsps mixed vegetable dice (scalded)
蒜蓉 2 茶匙	2 tsps minced garlic

調味料 | Seasonings

茄汁 1 湯匙	1 tbsp ketchup
黑胡椒碎 1 茶匙	1 tsp ground black pepper
糖 1 茶匙	1 tsp sugar
鹽 1/2 茶匙	1/2 tsp salt

做法 | Method

1. 熱鑊下油,爆香蒜蓉,放入茄汁意大利粉材料炒勻,再加入調味料拌勻。

2. 酥皮碾薄,厚約 3 毫米,按模具尺碼切成 2 塊,其中一塊放模具上定型,用叉刺孔,另一塊留作蓋面用。

3. 焗爐預熱 2-3 分鐘,先把已放模具的酥皮焗 5-8 分鐘,取出,再放上茄汁意大利粉,然後蓋上另一塊酥皮,掃上蛋液後刺孔,回爐焗至金黃即成。

1. Add oil to a heated wok, sauté minced garlic until fragrant, add ingredients of ketchup pasta and stir-fry well, add seasonings and mix well.

2. Roll pastry on a table into about 3 mm thickness, cut into 2 pieces according to the size of the mould. Line 1 piece onto the mould to shape and prick holes on it. While another piece reserved for covering the surface.

3. Preheat oven for 2-3 minutes, put pastry in the mould in oven and bake for 5-8 minutes. Take out and add ketchup pasta into the mould of pastry, then cover it with another piece of pastry. Brush with whisked egg and prick holes onto it. Bake in oven again until golden.

入廚貼士 | Cooking Tips

在酥皮上刺孔,可避免酥皮脹大時不平均。

Prick holes onto the pastry can prevent the pastry from swelling unevenly.

Baked Salmon and Penne Pasta
with Thousand Island Dressing

千島醬汁焗
三文魚長通粉

1~2 人
Serves 1~2

10 分鐘
10 minutes

材料 | Ingredients

即食長通粉 1 包　　　　1 pack instant penne pasta
三文魚（鮭魚）200 克　　200g salmon
蒙沙拉芝士 100 克　　　　3 tbsps thousand island dressing
千島醬汁 3 湯匙　　　　　100g mozzarella cheese

調味料 | Seasonings

雜香草 1 茶匙　　　　　1 tsp mixed herbs
黑胡椒碎適量　　　　　Pinch of ground black pepper

做法 | Method

1. 即食長通粉按包裝指示烹煮。
2. 三文魚切件，與千島醬汁拌勻後放入焗碟，再與長通粉和調味料撈勻，上面放芝士。
3. 焗爐預熱 2-3 分鐘，把長通粉放入焗爐焗 10 分鐘至表面呈金黃色，即成。

1. Cook instant penne pasta according to instructions on package.
2. Slice salmon, mix well with thousand island dressing and put in a baking plate, stir well with penne pasta and seasonings. Top with mozzarella cheese.
3. Preheat oven for 2-3 minutes, put penne pasta into oven and bake for 10 minutes until golden, serve.

入廚貼士 | Cooking Tips
三文魚可改用罐頭三文魚或吞拿魚。
Salmon can be replaced by canned salmon or tuna.

雜菌飯番茄盅

Tomato Pot with Assorted Mushroom Rice

◯◯◯ 材料 | Ingredients

鮮雜菌 50 克
番茄 1-2 個
雜色飯 1/2 碗
松子仁 1-2 湯匙
煙肉 2 片
薄荷 1-2 片（裝飾）

50g assorted fresh mushrooms
1-2 tomatoes
1/2 bowl mixed rice
1-2 tbsp pine nuts
2 slices bacon
1-2pcs mint (for dressing)

 調味料 | Seasonings

日式燒汁 1 湯匙　　　　1 tbsp Japanese sauce
鹽 1/2 茶匙　　　　　　1/2 tsp salt
糖 1/2 茶匙　　　　　　1/2 tsp sugar

做法 | Method

1. 煙肉切碎；雜菌洗淨，切碎。松子仁用白鑊烘香。
2. 燒熱油鑊，炒香煙肉，加入雜菌粒、雜色飯和松子仁炒勻，加入調味拌勻。
3. 番茄挖空，放入炒飯。
4. 焗爐預熱 2-3 分鐘，放入番茄盅焗 10 分鐘即成。

1. Mince bacon; rinse assorted mushrooms and mince. Toast pine nuts in a wok without adding oil.
2. Stir-fry bacon in a heated wok until fragrant, add assorted mushroom dice, colorful rice and pine nuts, stir-fry well, add seasonings and mix well.
3. Scoop out tomato flesh and put fried rice into it.
4. Preheat oven for 2-3 minutes, bake tomatoes in oven for 10 minutes, serve.

入廚貼士 | Cooking Tips

- 不用番茄可改用三色椒，效果相若但味道不同。
- Tomatoes can be replaced by green, yellow and red peppers, effect is similar but taste is different.

焗南瓜飯魷魚筒

Baked Whole Squid
with Pumpkin Rice

1~2 人
Serves 1~2

20~25 分鐘
20~25 minutes

材料 | Ingredients

鮮魷魚 1 隻	1 fresh squid
南瓜 200 克	200g pumpkin
即食日本紅豆飯 1/2 碗	1/2 bowl instant Japanese red bean rice
忌廉 50 毫升	50 ml cream

調味料 | Seasonings

鹽 1/4 茶匙	1/4 tsp salt
胡椒粉適量	Pinch of pepper

做法 | Method

1. 南瓜與清水同置煲中煮至軟腍，加入調味料和忌廉拌勻成濃稠狀。
2. 紅豆飯與南瓜蓉拌勻，待凍。
3. 魷魚洗淨，保留原隻。燒熱一鍋水，下魷魚焯至八成熟，釀入南瓜紅豆飯。
4. 焗爐預熱 2-3 分鐘，放入魷魚飯筒焗 10 分鐘，即成。

1. Put pumpkin and water into a pot together, cook until mushy. Add seasonings and cream and stir well to thicken.
2. Mix red bean rice and pumpkin mush well, leave to cool.
3. Rinse squid and keep in whole shape. Heat a wok of water and scald squid to 80% done. Stuff squid with pumpkin and red bean rice.
4. Preheat oven for 2-3 minutes, bake squid with rice in oven for 10 minutes, serve.

入廚貼士 | Cooking Tips

- 釀飯時必須壓實一點，較容易切開。
- Press hard when stuffing rice will make cutting open easier.

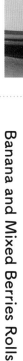

香蕉雜莓卷

Banana and Mixed Berries Rolls

材料 | Ingredients

酥皮 1 包	1 pack pastry
雞蛋 1 隻	1 egg
（打勻，掃面）	(whisked for surface brushing)

香蕉雜莓醬 | Banana and mixed berries paste

香蕉 1 隻	1 banana
雜莓（藍莓、士多啤梨和黑莓）100 克	100g mixed berries (blueberries, strawberries and blackberries)
糖 20 克	20g sugar
水 100 毫升	100 ml water

芡汁 | Thickening

粟粉 1 湯匙	1 tbsp cornstarch
水 2 湯匙	2 tbsps water

做法 | Method

1. 香蕉壓成果泥，備用。
2. 雜莓、糖和清水同置煲中以中火煮 5 分鐘，加入芡汁煮至濃稠，再加入香蕉拌勻。
3. 酥皮放軟，放枱上用木棒碾成厚約 5 毫米的長方形餅皮，掃上香蕉雜莓醬，捲好，切件，放焗盤上。
4. 焗爐預熱 3-5 分鐘，放入酥皮卷，表面掃上蛋液，焗 10-12 分鐘。

1. Mash banana into purée, set aside.
2. Put mixed berries, sugar and water into a pot and cook over medium heat for 5 minutes. Add thickening and cook until thickens, then add banana and stir well.
3. Let pastry stand until soft, put onto a table and roll into a rectangle shape in about 5 mm thick. Spread banana mixed berries paste onto pastry, roll up and cut into pieces. Put onto the baking tray.
4. Preheat oven for 3-5 minutes. Brush pastry rolls with whisked egg and bake in oven for 10-12 minutes.

入廚貼士 | Cooking Tips

- 如果酥表面或底部的顏色很深，可關掉上火或下火，否則可能會燒焦。
- If the color of the surface or bottom of the pastry turns pretty dark, switch off the upper fire or lower fire, otherwise the pastry will get scorched.

忌廉脆筒

Crispy Roll with Cream

2~4 人
Serves 2~4

12~15 分鐘
12~15 minutes

材料 | Ingredients

酥皮 1 包	1 pack puff pastry
雞蛋 1 隻（掃面）	1 egg (for surface brushing)
甜忌廉 100 毫升	100 ml sweet cream
粗砂糖適量	Some coarse sugar

做法 | Method

1. 雞蛋打勻；甜忌廉打起，備用。
2. 酥皮放軟，放枱上用木棒碾成厚約 5 毫米、長 8 厘米 × 闊 2 厘米的餅皮，捲在牛角模具。
3. 把牛角筒掃蛋液和灑上粗砂糖，放在已墊牛油紙的焗盤上。
4. 焗爐預熱 3-5 分鐘，放入牛角筒焗 10-12 分鐘或至鬆脆，取出待涼凍，唧上忌廉。

1. Whisk egg and whip sweet cream respectively, set aside.
2. Let pastry stand until soft, put onto a table and roll into about 5 mm thick and rectangular shape of 8cm x 2cm. Wrap onto the mould of horn.
3. Brush horn rolls with whisked egg and sprinkle with coarse sugar. Put onto baking tray lined with greaseproof paper.
4. Preheat oven for 3-5 minutes, bake horn rolls for 10-12 minutes or until crispy, take out and cool, add cream.

入廚貼士 | Cooking Tips

- 牛角筒表面開始出現燒焦現象時，可用錫紙蓋在上面以遮擋面火，卻不能熄掉，因為焗酥皮的火力不能太慢。
- If the surface of the horn rolls seem to be burnt, cover with a piece of aluminum foil to protect them from the fire but the fire cannot be switched off. It is because the heat cannot be too weak for baking pastry.

杏仁脆圈

Almond Crispy Rings

⟨⟨⟩⟩ 材料 | Ingredients

雞蛋 1 隻	1 egg
低筋麵粉 130 克	130g low gluten flour
牛油 90 克	90g butter
糖 60 克	60g sugar
杏仁粉 20 克	20g almond powder

⟨⟨⟩⟩ 裝飾 | Decoration

糖粉 100 克	100g castor sugar
蛋白 1 茶匙	1 tsp egg white
食用色素少許	Some edible colouring
杏仁粒適量（烘香）	Some almond grains (toasted)

做法 | Method

1. 牛油和糖打放碗中打至淡白色,加入雞蛋、低筋麵粉和杏仁粉拌勻成粉糰,裹上保鮮紙,待 10-15 分鐘。
2. 把粉糰分成 15-20 克的小粉糰,搓揉成橢圓形,中空如水泡形狀,放在已墊焗餅紙的焗盤上,放回冰箱冷凍 5-10 分鐘。
3. 焗爐預熱 3-5 分鐘,放入曲奇,用 190℃焗 15-20 分鐘。
4. 蛋白與糖粉拌勻,加入色素拌勻,塗抹在曲奇,黏上杏仁粒即可。

1. Whip butter and sugar in a bowl till pale white, add egg white, low gluten flour and almond powder and stir well into a dough. Wrap with a cling wrap and wait for 10-15 minutes.
2. Divide the dough into small pieces of dough of 15-20g each, rub into flat and oval shape with hollow in the middle (ring-like shape). Put onto a baking tray lined with parchment paper, put into the freezer to cool for 5-10 minutes.
3. Preheat oven for 3-5 minutes, put in cookies and bake at 190℃ for15-20 minutes.
4. Stir egg white and castor sugar well, add colouring and stir well, brush onto cookies and top with stick almond grains.

入廚貼士 | Cooking Tips

- 可用兩段火力烘烤曲奇,先用 190℃焗 10-15 分鐘,再轉用 160℃ -170℃慢火烘乾,效果會更好。
- Two kinds of heat can be used to bake cookies. Use 190 ℃ for 10-15 minutes first, then switch to 160 ℃ -170 ℃ low heat to bake for drying. Better effect will be obtained.

QQ 芝士球

QQ Cheese Balls

2~4 人
Serves 2~4

12~15 分鐘
12~15 minutes

材料 | Ingredients

QQ 粉 125 克	125g QQ powder
牛油 30 克（放軟）	30g butter (let stand until soft)
芝士粉 25 克	25g cheese powder
菜油 12 克	12g vegetable oil
雞蛋 12 克	12g egg
清水 107 毫升	107ml water

做法 | Method

1. QQ 粉、菜油和牛油放碗中拌勻。
2. 加入雞蛋和清水攪至成粉糰，再放進芝士粉拌勻，放擠袋中，唧在焗盤上。
3. 焗爐預熱，放入 QQ 芝士球，用上火 200℃、下火 160℃焗 10-20 分鐘即成。

1. Stir QQ powder, vegetable oil and butter well in a bowl.
2. Add egg and water to dough and mix well. Then add cheese powder and stir well. Put into squeezing bag and squeeze onto a baking tray.
3. Preheat oven, bake QQ cheese balls at 200℃ upper fire and 160℃ lower fire for 10-20 minutes, serve.

入廚貼士 | Cooking Tips

- 粉糰的攪打程序是成功的關鍵。攪麵糰時應先用慢速攪 1 分鐘，再下雞蛋和清水攪 1 分鐘，然後用中速攪 1 1/2 分鐘。
- The stirring procedure of dough is the key to success. When stirring dough, first stir by low speed for 1 minute, then add egg and water and stir for 1 minute, finally stir by medium speed for 1 1/2 minutes.

杏仁朱古力醬脆餅

Chocolate Cracker with Almond

材料 | Ingredients

蛋黃 1/2 隻
牛油 60 克（放軟）
低筋麵粉 60 克
高筋麵粉 60 克
糖 20 克
發粉（泡打粉）1/4 茶匙
原粒杏仁適量（烘香）

1/2 egg yolk
60g butter (let stand until soft)
60g low gluten flour
60g high gluten flour
20g sugar
1/4 tsp raising flour (baking powder)
Some whole grains of almond (toasted)

Dessert

(((朱古力醬 │ Chocolate saue

朱古力 100 克
淡忌廉 10 克
牛油 5 克

100g chocolate
10g whipping cream
5g butter

(((做法 │ Method

1. 朱古力與淡忌廉放碗中,座熱水上拌勻至融,加入牛油拌勻,備用。
2. 牛油與糖放碗中打至鬆軟,加入蛋液拌勻。
3. 把粉材料一同過篩,混入牛油混合物內拌勻,搓揉成粉糰。
4. 工作枱上鋪上牛油紙,放上粉糰,用木棒碾成 3 毫米厚,置冰箱冷凍 30 分鐘,取出壓成圓形,其中 1 片中空。將每片中空的脆餅放在圓形脆餅上。
5. 焗爐預熱 3-5 分鐘,放入脆餅,用 170℃焗 15-20 分鐘,待涼,倒入朱古力醬,放上杏仁裝飾便可。

1. Put chocolate and whipping cream in a bowl and place the bowl over hot water, stir till melt, add butter and stir well, set aside.
2. Whip butter and sugar in a bowl until soft, add whisked egg, stir well.
3. Sieve flour ingredients together, mix with butter mixture, rub into a dough.
4. Put the dough onto the table lined with greaseproof paper, roll the dough to 3 mm thick. Put into freeze for 30 minutes, take out and mould into circular shape, make hollow in the middle in one of the pieces. Place the piece of cracker with hollow onto the round cracker.
5. Preheat oven for 3-5 minutes, bake crackers at 170℃ for 15-20 minutes. Leave to cool, pour in chocolate sauce and add almond for decoration.

芝士麵包條

Cheese Bread Sticks

4~5 人
Serves 4~5

15~20 分鐘
15~20 minutes

 材料 | Ingredients

高筋麵粉 100 克	100g high gluten flour
低筋麵粉 70 克	70g low gluten flour
芝士粉 30 克	30g cheese powder
乾酵母 3 克	3g dry yeast
糖 3 克	3g sugar
鹽 2 克	2g salt
雞蛋 1 隻（掃面）	1 egg (for surface brushing)
清水 110 毫升	110ml water

做法 | Method

1. 把乾酵母、糖和清水同置碗中靜待 2 分鐘。
2. 將所有粉材料一同過篩，倒入酵母水，搓揉成粉糰，裹上保鮮紙，靜置 1 小時。
3. 粉糰分成 20 份，搓幼成條狀，置焗盤上靜待發酵至約 2 倍大。
4. 焗爐預熱 3-5 分鐘，以 160℃ 焗 20 分鐘或至脆硬，呈金黃色便可。

1. Put dry yeast, sugar and water into a bowl together, wait for 2 minutes.
2. Sieve all flour materials together into the yeast water, rub into a dough. Wrap with cling wrap and set aside for 1 hour.
3. Divide dough into 20 pieces and rub into thin stick shape. Put onto baking tray and let ferment till it expands double in size.
4. Preheat oven for 3-5 minutes, bake at 160 ℃ for 20 minutes or until crispy and golden brown, serve.

入廚貼士 | Cooking Tips

- 麵糰無論搓好或已造型，必須用保鮮紙蓋面，防止表面風乾。
- No matter the dough is shaped or not, it must be covered with cling wrap to prevent the surface from drying.

Pumpkin Muffins

南瓜鬆餅

 材料 | Ingredients

南瓜 80 克	80g pumpkin
高筋麵粉 70 克	70g high gluten flour
低筋麵粉 60 克	60g low gluten flour
牛油 60 克	60g butter
糖 60 克	60g sugar
鮮奶 50 毫升	50ml milk
發粉（即泡打粉）1/2 茶匙	1/2 tsp raising flour(baking powder)
雞蛋 1 隻	1 egg

做法 | Method

1. 南瓜洗淨，去皮去核，以大火隔水蒸熟，壓蓉備用。

2. 牛油與糖放碗中拌至乳白色，加入雞蛋拌勻，倒入鮮奶和南瓜蓉拌勻。

3. 各式粉材料一同篩勻，加入南瓜蓉混合物拌勻，倒入多個紙杯內。

4. 焗爐預熱 3-5 分鐘，放入鬆餅，用 160℃ 焗 15-20 分鐘，即成。

1. Rinse pumpkin, peel and seed. Steam pumpkin till done over high heat, mash into purée, set aside.
2. Stir butter and sugar in a bowl till milky white, add egg and stir well. Pour in milk and pumpkin mash and stir well.
3. Sieve flour ingredients together into pumpkin mash mixture, stir well, pour into paper cups.
4. Preheat oven for 3-5 minutes, bake muffins at 160℃ for 15-20 minutes, serve.

入廚貼士 | Cooking Tips

- 已攪拌的粉漿必須盡快入爐，否則會影響到製品效果。
- The already stirred flour paste need to bake as soon as possible, otherwise the result will be affected.

焗西米布甸

Baked Sago Pudding

4~6 人
Serves 4~6

12~15 分鐘
12~15 minutes

 材料 | Ingredients

糖 80 克	80g sugar
西米 60 克	60g sago
牛油 40 克	40g butter
鮮奶 25 毫升	25ml milk
吉士粉 20 克	20g custard powder
粟粉 20 克	20g cornstarch
蛋黃 2 隻	2 egg yolks
水 450 毫升	450ml water

做法 | Method

1. 西米浸軟，瀝乾，備用。

2. 把蛋黃、鮮奶、吉士粉和粟粉放碗中拌勻成混合物。

3. 清水放鍋中煮滾，加入混合物一同煮至熟透，倒入已塗牛油熔液的小焗盅。

4. 焗爐預熱 3-5 分鐘，放入西米布甸，以 190℃焗 10-15 分鐘，即成。

1. Soak sago till soft, drain and set aside.

2. Stir egg yolk, milk, custard powder and cornstarch in a bowl well into a mixture.

3. Boil water in a pot, add mixture and bring to a boil. Spread melted butter onto baking pots and pour in mixture.

4. Preheat oven for 3-5 minutes, bake sago pudding at 190℃ for 10-15 minutes, serve.

入廚貼士 | Cooking Tips

- 可加入蓮蓉或豆沙作夾層餡料。
- You may add lotus seed paste or red bean paste as filling.

焦糖布甸

Caramelized Pudding

材料 | Ingredients

鮮奶 350 毫升
鮮忌廉 80 克
糖 80 克
雞蛋 4 隻（打勻）

350 ml milk
80g fresh cream
80g sugar
4 eggs (whisked)

焦糖 ｜ Caramel

糖 150 克   150g sugar
熱水少許   Some hot water

做法 ｜ Method

1. 燒熱鍋，將焦糖中的糖放入鑊中，以慢火煮至融化，開始轉變成棕色時，加入熱水繼續煮至溫度達 105℃，備用。
2. 燒熱鍋，加入鮮奶、鮮忌廉和糖，以慢火加熱，拌勻至糖融化，但不能煮至沸騰。
3. 熄火，加入蛋液拌勻，以隔篩過濾。焦糖放在焗盅內，加入布甸混合物。
4. 焗爐預熱 3-5 分鐘，把焗盅放進焗盤，注水在焗盤，用 150℃焗 35-45 分鐘至凝固。

1. Melt sugar (for caramel) in a wok over low heat, add hot water when the color begins to turn brown, continue to add hot water and cook until the temperature reaches 105℃ .
2. Heat a wok, add milk, cream and sugar together and stir until sugar melts over low heat, do not bring to a boil.
3. Switch off heat, add whisked egg and stir well, filter. Put caramel in baking pots and then add pudding mixture.
4. Preheat oven for 3-5 minutes and put baking pots onto a baking tray. Add water to baking tray and bake at 150℃ for 35-45 minutes until firm.

入廚貼士｜Cooking Tips

- 因應焗爐的尺碼而決定烹調時間。
- Adjust cooking time according to oven size.

肉桂蘋果撻

Cinnamon Apple Tart

(OO) **撻底** | Bottom of the tart

低筋麵粉 150 克	150g low gluten flour
牛油 90 克（放軟）	90g butter（let stand until soft）
糖 45 克	45g sugar
雞蛋 1/2 隻	1/2 egg

(OO) **蘋果餡** | Apple fillings

青蘋果 1 個（切粒）	1 green apple (dice)
糖 15 克	15g sugar
肉桂粉 1 茶匙	1 tsp cinnamon powder
葡萄乾少許	Some raisins
核桃碎少許	Some ground walnuts

4~6 人
Serves 4~6

20~25 分鐘
20~25 minutes

掃面 | For surface brushing

雞蛋 1 隻（打勻） 1 egg (whisked)

做法 | Method

1. 燒熱鑊，下糖和蘋果粒煮片刻，待涼，放進蘋果餡的其他材料拌勻。
2. 牛油與糖放碗中拌勻，加入蛋漿拌勻後，與低筋麵粉拌勻，搓勻成粉糰，置冰箱中冷凍 15 分鐘。
3. 取出 1/2 份粉糰，放枱上，用木棒碾平厚約 2 毫米，放在撻模上，裁去多餘的部分，再放上蘋果餡。
4. 把其餘粉糰用木棒碾平，切成長條，鋪在撻面，掃上蛋液。
5. 焗爐預熱 3-5 分鐘，放入蘋果撻焗 20-25 分鐘，即成。

1. Heat a wok, add sugar and apple dice and cook for a while, leave to cool. Add other ingredients and stir well.
2. Stir butter and sugar in a bowl well, add whisked egg and stir well. Add low gluten flour, mix well and knead into a dough. Put into freezer for 15 minutes.
3. Take out half of the dough and roll flat to about 2 mm thick. Put onto tart mould and trim off excess dough, then add apple fillings.
4. Roll remaining dough flat and cut into long strips. Put on the top of the tart in pattern, brush with whisked egg.
5. Preheat oven for 3-5 minutes, bake apple tart for 20-25 minutes, serve.

入廚貼士 | Cooking Tips

- 撻面的裝飾可利用剩餘麵糰按自己喜好創作。
- You can make use of the remaining dough to decorate the tart's surface according to your preference.

焗爐美食

編著
方智敏

編輯
紫彤

美術設計
Venus Lo

排版
何秋雲

翻譯
Wong Hing

攝影
Wilson Wong

出版者
萬里機構出版有限公司
香港鰂魚涌英皇道1065號東達中心1305室
電話：2564 7511
傳真：2565 5539
電郵：info@wanlibk.com
網址：http://www.wanlibk.com
　　　http://www.facebook.com/wanlibk

發行者
香港聯合書刊物流有限公司
香港新界大埔汀麗路36號
中華商務印刷大廈3字樓
電話：2150 2100
傳真：2407 3062
電郵：info@suplogistics.com.hk

承印者
美雅印刷製本有限公司

出版日期
二零一八年十一月第一次印刷

萬里機構　　萬里 Facebook